For Juliette, always seeking new places to explore

The illustrations for this book were made with acrylic paint on paper.

Cataloging-in-Publication Data has been applied for and may be obtained from the Library of Congress.

ISBN 978-1-4197-5682-5

Text and illustrations © 2025 Marc Majewski
Book design by Pamela Notarantonio

Published in 2025 by Abrams Books for Young Readers, an imprint of ABRAMS. All rights reserved. No portion of this book may be reproduced, stored in a retrieval system, or transmitted in any form or by any means, mechanical, electronic, photocopying, recording, or otherwise, without written permission from the publisher.

Abrams® is a registered trademark of Harry N. Abrams, Inc.

Printed and bound in China
10 9 8 7 6 5 4 3 2 1

Abrams Books for Young Readers are available at special discounts when purchased in quantity for premiums and promotions as well as fundraising or educational use. Special editions can also be created to specification. For details, contact specialsales@abramsbooks.com or the address below.

ABRAMS The Art of Books
195 Broadway, New York, NY 10007
abramsbooks.com

PARKS

Marc Majewski

Abrams Books for Young Readers

New York

Parks are BIG

Parks are small

PARKS ARE LOUD

Parks fly

Parks float

Parks twinkle

Parks protect

Parks invent

Across the world or down the street...

ABOUT THE PARKS

PARKS ARE BIG
Central Park, USA

Central Park is a big park that spreads out over 1.3 square miles in the middle of Manhattan in New York City. It is even bigger than the whole country of Monaco! The park features a vast 55-acre lawn and walking paths that wind over 58 miles alongside many lakes. It's filled with over 18,000 trees, making it feel like a forest—one with tall buildings peeking out above the leaves.

PARKS ARE SMALL
Place Dauphine, France

Place Dauphine is a small square hidden on an island called Île de la Cité in Paris, France. Known for its triangle shape, the park is surrounded by old buildings with cafés and restaurants. Despite its modest dimensions, the park's unique design and intimate atmosphere make it one of the prettiest parks in the city.

PARKS ARE QUIET
Hampstead Heath, England

Hampstead Heath is one of the largest and most peaceful parks in London, England. With its dense woodlands, the park is home to many animals and over 180 bird species. It is also a cherished place for anyone who wants to get away from loud city noises and enjoy some quiet. In 2021, Hampstead Heath was even honored as the first Urban Quiet Park in Europe by a nonprofit organization called Quiet Parks International.

PARKS ARE LOUD
Tivoli Gardens, Denmark

Tivoli Gardens is a park located in the heart of Copenhagen, Denmark. It is one of the oldest amusement parks in the world, and its wooden roller coaster, built in 1914, is the oldest operating roller coaster in Europe. Around 4 million visitors come to Tivoli Gardens each year—even the Danish royal family! The park has over 20 rides, and it hosts many fun events including concerts, musicals, and fireworks shows.

PARKS ARE TIDY
Versailles Gardens, France

The Versailles Gardens, not far from Paris, France, were designed in 1661 by landscape architect André Le Nôtre; the park is one of the most famous examples of French formal gardens. This type of garden is characterized by a tidy and organized layout, symmetrical arrangements, precise pathways, and aligned rows of trees. The geometric design is meant to create an extraordinary sense of order and harmony.

PARKS ARE MESSY
Emdrup Junk Playground, Denmark

The Emdrup Junk Playground is an adventure playground that was built outside of Copenhagen, Denmark, in 1943. The concept of an "adventure playground" was developed in the 1930s by the landscape architect Carl Theodor Sørensen. It can be defined as a place designated for play that encourages kids to use recycled materials and everyday objects to build their own imaginative structures and games. In these junk parks, children can freely explore, experiment, and collaborate. Today, there are many adventure playgrounds in Europe and around the world, another of which is the Yard on Governors Island in New York City.

PARKS ARE SUNNY
Death Valley National Park, USA

Death Valley National Park covers a territory of 3,422,024 acres in the southwestern United States, located in California near the border to Nevada. It is famous for its extreme climatic conditions, holding the record for the hottest temperature ever recorded on Earth, with a scorching high of 134°F (56.7°C). The park has big areas of salty ground, tall sand hills, and rocky mountains. Despite its harsh reputation, some plants and animals have figured out how to live there, and many visitors come every day to admire its remarkable desert landscape.

PARKS ARE SNOWY
Northeast Greenland National Park, Greenland

Northeast Greenland National Park is the world's largest national park, covering an area of approximately 375,000 square miles. With extremely low temperatures and tough winters, it is one of the coldest places on the planet. The park is mainly made up of glaciers, ice caps, and rugged mountains, where Arctic animals like polar bears, musk oxen, and walruses live. This park is also important for studying climate change in the Arctic and preserving the world's ice sheets.

PARKS FLY
High Line, USA

The High Line in New York City, USA, is an urban park built on an abandoned elevated railway. The old tracks have been reimagined as a living ecosystem, integrating architecture and ecology. Spanning 1.45 miles across different neighborhoods of Manhattan, it offers New Yorkers an exceptional view of the city and the Hudson River as it seemingly flies 30 feet above the busy streets. The park is filled with various plants, trees, and grasses that provide habitats for many species like birds, bees, and butterflies.

PARKS FLOAT
Keibul Lamjao National Park, India

Keibul Lamjao National Park, located in the northeastern state of Manipur, India, is the only floating national park in the world. The park spreads out over 15.4 square miles and is famous for its unique floating vegetation, called Phumdis, which are essentially masses of organic material, soil, and plants that float on the surface of the Loktak Lake. Local fishermen often use traditional wooden boats to move around these floating islands, which are home to many animals, including the very rare and critically endangered Sangai deer.

PARKS TWINKLE
Gabriela Mistral Dark Sky Sanctuary, Chile

The Gabriela Mistral Dark Sky Sanctuary, named after a famous Chilean poet, is a perfect place for stargazing and astronomical observations. Due to its location far from cities, this park in northern Chile is protected from any light pollution, offering visitors some of the darkest and clearest skies on Earth and a wonderful occasion to contemplate the Milky Way, bright stars, meteor showers, and distant galaxies.

PARKS DAZZLE
Keukenhof Gardens, Netherlands

Keukenhof Gardens in Lisse, also called the "Garden of Europe," is the most famous garden in the Netherlands and one of the biggest flower gardens in the world. It is filled with millions of colorful flowers like tulips, daffodils, and hyacinths. Each autumn, gardeners plant over 7 million bulbs, including 800 different varieties of tulips. The park is open only in spring, inviting visitors to enjoy the many blooming flower fields.

PARKS PLAY
Parque Francisco Alvarado, Costa Rica

Parque Francisco Alvarado is a topiary park in the town of Zarcero, Costa Rica. It was taken over in the 1960s by gardener and artist Evangelisto Blanco, who playfully trimmed the hedges into a variety of shapes and figures, including animals and abstract forms. "Topiary" refers to the gardening practice of trimming and shaping trees and shrubs into decorative shapes. Visitors can wander the park's pathways to admire the various plant sculptures, which create a unique blend of nature and art.

AND PLAY
Isamu Noguchi's Piedmont Park Playscapes, USA

Isamu Noguchi was a famous American artist and landscape architect who designed many sculptures, public spaces, and pieces of furniture throughout his career. In 1975, he was commissioned to create a playground in Piedmont Park in Atlanta, Georgia, that would serve as both a work of art with distinctive geometric sculptures and an interactive space for children. He called it Playscape. During his life, Noguchi designed and created many Playscapes, as he believed that "everything is sculpture" and that art should be integrated into people's everyday lives.

AND PLAY SOME MORE
Nishi-Rokugō Park, Japan

Nishi-Rokugō Park, also called "tire park," is a playground in Tokyo, Japan, filled with around 3,000 recycled tires of various sizes. Some of them are piled up to form big monsters and robots; some are turned into swings or bridges. The park is made for children to explore, climb, and play. Similar to adventure playgrounds, the Nishi-Rokugō Park is a perfect example of how waste can be reimagined into something fun.

PARKS BARK
Ohlone Dog Park, USA

Opened in 1979 in Berkeley, California, the Ohlone Dog Park is recognized as the first dog park in the United States. It's a spacious green area where dogs can run and play without leashes. Dog parks are designed for dogs to exercise and play in a safe environment, and they are typically divided into sections for both big and small dogs.

PARKS ROAR
Serengeti National Park, Tanzania

Serengeti National Park, located in Tanzania, is a vast park stretching over 5,700 square miles and teeming with wildlife. The park is known for a yearly event called the "Great Migration," during which about 2 million wildebeests, zebras, and other animals move across the land in search of water and food. The park is home to lots of different animals, including elephants, buffalo, leopards, rhinoceroses, and lions. Lions are especially famous there; more lions live in the Serengeti than in any other place in Africa.

PARKS PROTECT
Great Barrier Reef Marine Park, Australia

The Great Barrier Reef Marine Park is part of the Great Barrier Reef, the largest coral reef system in the world, off the coast of Queensland, Australia. The park, stretching over 132,806 square miles, is home to incredibly rich biodiversity with countless species of coral, fish, and marine animals. The mission of the park is to protect this precious ecosystem from being damaged by human activities, such as intensive fishing and pollution. The park is also an important space for scientists and researchers studying marine biology, oceanography, and climate change.

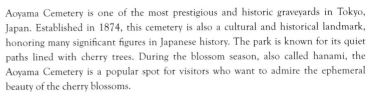

PARKS HONOR
Aoyama Cemetery, Japan

Aoyama Cemetery is one of the most prestigious and historic graveyards in Tokyo, Japan. Established in 1874, this cemetery is also a cultural and historical landmark, honoring many significant figures in Japanese history. The park is known for its quiet paths lined with cherry trees. During the blossom season, also called hanami, the Aoyama Cemetery is a popular spot for visitors who want to admire the ephemeral beauty of the cherry blossoms.

PARKS INVENT
Gardens by the Bay, Singapore

Gardens by the Bay is a futuristic park that covers 250 acres in the heart of Singapore. It is famous for various attractions like the Flower Dome—the largest glass greenhouse in the world!—and the Supertree Grove. Supertrees are man-made vertical gardens that look like huge trees, reaching up to about 160 feet tall and filled with more than 162,900 plants from over 200 species. The Supertrees are also built to be eco-friendly: They use solar panels to get energy from the sun and have systems to collect rainwater. This mix of technology with plants shows an inventive way to make cities greener and more sustainable.

PARKS ENCHANT
Las Pozas, Mexico

Las Pozas is a Surrealist sculpture garden located near the village of Xilitla in the Sierra Gorda, a mountain range in Mexico. Created by British poet and artist Edward James in the mid-twentieth century, this enchanting place spreads over 80 acres in the middle of the jungle. It is filled with whimsical sculptures, enigmatic concrete structures, and winding staircases that seemingly lead to nowhere. These sculptures have dreamlike names, such as *The House with a Roof like a Whale* and *The Staircase to Heaven*. As a patron of the Surrealist art movement, James was inspired to create a "Garden of Eden," and he dedicated much of his fortune and the later part of his life to developing this garden, which is now one of the region's most famous and unique attractions.

Other Parks Shown

**JACKET, COVER, "ACROSS THE WORLD..."
AND "PARKS WELCOME":**
My inventions!

TITLE PAGE:
Inspired by the Emdrup Junk Playground
(from "PARKS ARE MESSY")